A Wisley Handbook

Fragrant and Aromatic Plants

KAY N. S

Ca

The Royal Horticultural Society

THE ROYAL HORTICULTURAL SOCIETY

Cassell Educational Limited
Villiers House, 41/47 Strand
London WC2N 5JE
for the Royal Horticultural Society

First edition 1985
Reprinted 1987
Second edition 1993

British Library Cataloguing in Publication Data
A catalogue record for this book is available from the British Library

ISBN 0-304-32033-1

Photographs by Pat Brindley, Iris Hardwick Library of Photographs, Photos Horticultural and Paul Roberts

Phototypesetting by RGM Associates, Southport
Printed in Hong Kong by Wing King Tong Co. Ltd

Cover: syringa and wisteria
Photograph by Andrew Lawson
Frontispiece: *Acacia dealbata*
Back cover: *Iris unguicularis*
Photographs by Iris Hardwick

Contents

Introduction	5
Herbaceous Perennials and Annuals	9
Trees and Shrubs	25
Bulbs and Corms	43
Aromatic Plants	53
Further Reading	63
Index	64

Introduction

There is no need to understand the formation and composition of scent to be able to plant and enjoy a fragrant garden. The scent and aroma of flowers and leaves are due to their essential oils. The latter term is quite misleading because, far from being essential, the oils are considered to be by-products of the plant, which being volatile react with the atmosphere to produce a varied array of scents. Because of the complex chemical composition of these oils and attars, the tiniest variation can change the scent of the end product. Variations occur during the season, or simply through the life of the plant, and the scents are usually at their richest when the plant is fully mature and just as the flowers are about to open. Some plants, of course, only become scented as they fade, or even after drying, and the various nuances of the character of a basic scent add much interest to the fragrant garden through the season.

Fragrance is the term usually applied to flower scent and those scents that are given off freely by plants. Aroma always seems to suggest the scents that are locked away within the plant and which need to be released (for example by crushing the foliage) to be evident. The quality of a scent is a highly personal appreciation, consequently a curiously inadequate vocabulary has evolved to describe the plant scents. One plant is often used to name the scent of another: violet-scented, pineapple-scented, almondy, minty or musky. Scent gives the garden a fourth dimension, and this book draws attention to some of the scented plants that are readily available and suggests ways in which they may be used to advantage. Since details of cultivation and descriptions of the plants are available elsewhere, the space here is used primarily to consider the quality of scent.

There are many simple ways in which the enjoyment of fragrance can be heightened, the keynote being shelter. The first essential is to provide or select a site sheltered from prevailing winds, and ideally endowed with a *light* canopy of trees to help to ensnare the fragrance rather than allow it to be snatched by the wind. Various 'tricks' can be devised to render fragrance more accessible, such as planting in raised beds or troughs, or using mobility by growing

---

Myrtus communis, the common myrtle (see p. 33), flowers in May and June

Roses and lavender look good together and smell delicious

plants in tubs or boxes that can be moved from place to place. Window boxes also bring scent to where it can be enjoyed, especially by the housebound. Another idea is to plant against a warm wall where the radiated heat from the brickwork or stonework will enhance evening scent and help to lengthen the period of enjoyment. It is a remarkable fact that the supply of essential oils, and therefore scent, from a plant seems to be inexhaustible, and is often related to the weather, warmth encouraging more scent. Pathways can be dotted with aromatic plants that tolerate being trodden on, or lined with sub-shrubs that when brushed in passing donate their contribution to the enjoyment of the garden.

Holding aloft is the major way in which scented plants can be enjoyed. Space in the garden may not allow for a pergola, but simple canopies like car ports, or patio arbours, can support scented climbers. Trellis formed of, or covered by, 'Netlon'-style mesh can form bays in borders, nose-high fences, screens, arches or columns to support various scent-producing plants. A simple tripod or 'wigwam' of canes, strategically placed, can be a decorative feature and at the same time allow air to pick up the scent. Old tree stumps, dilapidated sheds and disused seats all offer support for fragrant scramblers to good garden effect.

Consider the evening-scented qualities of plants in gardens where the owners are away at work during the day, or use the prevailing breeze to waft a scent towards the windows of the house or to the patio. The season of flowering or garden effect obviously has to be taken into consideration as with any other garden planning, and while care needs to be exercised in not masking a delicate scent by an overpowering one, in practice it hardly ever happens that scents clash and become unacceptable.

Catmint, *Nepeta* × *faassenii*, makes a good edging plant for paths (see p. 17)

Herbaceous Perennials and Annuals

In the decorative garden it is the flowering plants that provide the blocks of colour, varying from one month to the next. Relatively few of them are richly scented although these are the ones we gather to form a true nosegay. Many of them are hardy perennials and because they are raised and managed easily they are popular, and are often the first plants the amateur attempts to grow. Flowering plants are usually grouped for colour effect and form in borders, or on the edge of shrub borders; nowhere do they look better than when massed among others of their kind. The contrast of one colour or leaf form, or even plant form, with the next is part of garden artistry.

In the border, the year starts with the bergenias, or elephant ears, so sweetly scented, the more so when taken indoors, and soon after come the erysimums, closely related to the wallflower. Then follow irises and peonies, *Crambe cordifolia* for the back of the border or a rough area, achilleas and their relative, tansy. Fragrant annuals to try include *Lobularia maritima*, available with flowers in rich pinks as well as white, nasturtium (*Tropaeolum majus*), marigold (*Calendula officinalis*, see p. 10), French and African marigolds (*Tagetes*) and the sweetest scented of all, the sweet pea (*Lathyrus odoratus*). Present-day dwarf varieties of sweet pea are deliciously scented and at the same time they are labour-saving and can be used to good effect as border edges or to line a path where the fragrance can be enjoyed.

The tree lupin (*Lupinus arboreus*) is easily raised from seed and lingers for a few years in many areas, forming a shrubby base and sprouting fragrant yellow flowers. It is soon established and therefore useful for quick effect in a new garden.

***Cheiranthus cheiri*, the wallflower,** is normally cultivated as a biennial; seed is sown one spring to flower in the following year (see p. 10). Few plants are as hardy and tolerant of any and every situation, be it open border, or container, and utterly reliable for a good spring show. Their perfume is among the richest of all flower perfumes, riding on the air even from one garden to the next, when

---

Heliotropium arborescens (*H. peruvianum*), cherry pie, is an almond-scented half hardy annual (see p. 14)

Above: Wallflowers, *Cheiranthus cheiri*
Below: Pot marigold, *Calendula officinalis*

planted in a mass. Single or double velvety flowers come in a wide range of colour from cream to deep maroon and brilliant orange, crimson and purple. Best results are achieved when seed is sown early in spring and plants transplanted to permanent quarters in late September or October. Sun or semi-shade seem to make little difference to the results, but they respond to good conditions. Do not feed after planting.

***Convallaria majalis*, lily-of-the-valley,** (see p. 13), produces a scent sweeter than any other plant. Native of our woodlands and therefore thriving best in leafmouldy, moist situations, it has dainty clear white rounded bells dangling along one side of the stiff stem. The foliage pierces the ground first like a little rolled umbrella before opening into twin broad flat leaves in early May. Choose shade or semi-shade beneath trees or shrubs or in woodland or on a north-facing bank for the most satisfactory results, and heed a predeliction for chalky soil. The perfume persists, even after cutting, until the flowers fade. Specially prepared crowns are available for gentle forcing in pots, when the flowers will bloom in the conservatory or porch in March and April.

***Dianthus*, carnations and pinks or gillyflowers** Among the oldest of cultivated plants, there is no doubt that they have been cherished for their delicious fragrance. Always spicy and full bodied, some scents display distinct clove-like overtones – lending the name, for example, to 'clove pink'. They flourish on chalk and almost bare limestone to produce dainty flat-faced little flowers held at right angles to the stem and set off by a tuft of grey-green thick grass-like leaves. Various sorts are useful for the rock garden, tucked into wall pockets, for pots, troughs or open border or as conservatory decoration. Most can be grown from seed, and there is an almost bewildering choice of named cultivars, all of which carry the distinctive scent. Some are annual, many perennial, and the latter can be propagated from cuttings or layers or by division after flowering.

***Dianthus barbatus*, sweet william,** (see p. 13), stand apart from the remainder of the dianthus in that the flowers are held together in a broad flattened head, fringed by a green beard. They are proudly upstanding on tough bright green stems and offer a rich perfume, sweeter and less clove-toned than their relatives, the pinks. Generally, the auricula-eyed forms bear the sweetest scent. Sweet william is a short-lived perennial, and is often grown as a biennial, especially in the colder areas; cutting back after flowering will encourage the basal growth to thicken up and make a bolder plant and at the same time prolong life.

Dianthus caryophyllus is the species from which the hardy border carnations have been developed. Border carnations are short-lived perennials and not at their best in the mixed flower border. They have double frilled flowers often with notched petals and a rather nutmeg-clove scent deeper than that of the pinks, and bloom mainly in July and August. A strain intermediate between the border carnation and the perpetual-flowering (glasshouse) carnation is known as 'cottage' carnations. They have shorter stouter stems and the flowering season extends to September. Seedsmen's lists are worth perusing and novelties are worth trying, especially as an experiment to grow in a pot.

***Dianthus plumaris*, the pink,** no longer lives up to this name but comes in the most glorious colour range – especially among the hybrid allwoodii sorts. Hardy, pert, tolerant of drought, lime and clipping, pinks will flower over a long period from their peak in midsummer until well into September. Their faces are painted or plain and they are generally grouped according to the way in which the colour is disposed on the petals; self (one colour), bi-color, laced or fancy. The classification depends upon parentage, and as scent is transferred from parents to progeny, all groups have a place in the fragrant garden. The groups are garden pinks, allwoodii pinks, London pinks, and hybrid alpine pinks.

Garden pinks are mainly heavily scented and burst into flower once only during June. Burst is perhaps the most appropriate word because they all suffer from an inadequate calyx which splits, allowing the petals to spill out of formation. The well-known 'Mrs Sinkins' is an example.

Allwoodii pinks are distinctly clove-scented, with a neater habit, somewhat squat. Invaluable as border edging plants, they flower intermittently between June and August. Most bear Christian names, easily remembered when identified with one's acquaintances: 'Doris', 'Isobel', Monty' and 'Ian'.

London pinks are really a cross between the two previous groups (with some other intermarrying too) and have pretty flowers laced and edged with deeper colours usually matching the eye. Quite long in the stem for pinks, but not too lanky to be useful border plants, they are so-called because their names commemorate the capital: 'London Lovely' and 'London Poppet'.

Above: Lily-of-the-valley, *Convallaria majalis*, has one of the most favoured scents (see p. 11)
Below: *Dianthus barbatus*, sweet william (see p. 11)

Hybrid alpine pinks are those small plants that are perfect for a trough on the patio, or for growing among paving or in walls. Their small flowers bloom over a long period in June and July, scenting the air for all they are worth. The real alpine pinks originate from *D. alpinus* and are diminutive plants, smothered in tiny flowers throughout the spring and summer and they are useful for the rock garden. Crosses have been made with them and *D.* × *allwoodii* and a packet of seed of Allwoodii Alpinus Mixed could produce a host of variations on the theme of little pink fringed faces with eye colours from the lipstick range.

***Filipendula ulmaria*, the meadowsweet or queen of the meadows,** used to be known as *Spiraea ulmaria*, and is still occasionally catalogued thus. Enjoy it for its deep, over-sweet almond fragrance, sometimes a little fishy during very dry weather. Frothy creamy white blossoms, held upright on good stems decorated with fern-like leaves, look particularly good at the margin of a pond or in damp areas. It prefers damp soils and will thrive in full sun or semi-shade. Track it down in its decorative-leaved forms of 'Aurea' or 'Variegata' to add an air of gaiety to the garden when it is in flower during July and August.

***Heliotropium arborescens* (*H. peruvianum*), cherry pie,** grown for its almost sickly almond scent, is best cultivated as a half hardy annual (see p. 8). Use it as a border to flower beds, as an edging to a path or in tubs and give it good drainage, for it hates to have wet roots. Pot up a few seedlings to grow in the conservatory or porch, where they will flower long after their companions used for summer bedding have been cleared. The leaves are an interesting deep green and many-veined and form a good backing to the clustered deep purple and white flowers. Various flower colours have been selected and are available from seedsmen; those paler in colour seem to be stronger in fragrance.

***Hesperis matronalis*, sweet rocket or dame's violet,** is one of the loveliest of cottage garden plants for dry limy soil. It is a biennial that once grown will seed itself about the garden and stay for many years to produce lilac and purple or sometimes white flowerheads on tall swaying stems in May. Double-flowered forms are worth seeking out and are considered to be more strongly scented. During the day its violet fragrance needs to be sought out but twilight strengthens it to a rich sweet scent with spicy overtones that spill out on to the surrounding air. A lovely evening garden plant which

Limnanthes douglasii, the poached egg plant

our ancestors used to call queen's gillyflower, because it was considered superior in scent to all the other gillyflowers, that is stocks, wallflowers and pinks.

***Limnanthes douglasii*, poached egg plant,** because the white daisy-like flowers are marked with a broad shiny yellow central disc. But that is the only resemblance to eggs; the scent is rich and sweet, especially in calm weather. Used as a border edging plant, scattered among paving stones or used to cover a bank, limnanthes produces a lovely effect of flowing foam. It will flower from May until late summer, attracting bees to the garden. Perfectly hardy in spite of being a Californian native, it will spread in areas where seed has been sown and stay for several years by re-seeding itself.

***Melilotus officinalis*, yellow melilot,** is probably shunned by the purists as a wild plant and weed, but it is a good nectar plant, producing lovely branches of sweet-smelling golden pea-like flowers at midsummer. Formerly a fodder plant and now creeping into herb gardens, it is worth a corner where the soil may be poor and the sun can light up its yellow flowers for at least half the day. Dry the sprigs to enjoy the fragrance as it increases and lasts. It is a biennial, and seed is available from seedsmen specialising in wild flowers. It is also a good bee plant.

***Monarda didyma*, bergamot or bee balm,** is a quick-growing perennial, which loves nothing better than a good supply of moisture-retentive material about its roots. Good clumps are formed, often deteriorating in the middle after three years or so; therefore constant division is the key to continued success. Flowers of the species are a deep glowing red arranged in whorls around the stem and there are cultivars with various coloured flowers, all retaining the scent: 'Croftway Pink', 'Snow Maiden', 'Prairie Night' and 'Hartswood Wine'. But the most popular is 'Cambridge Scarlet' which is a good red. The foliage holds much of the perfume, but the whole plant is impregnated and, even after top growth is cut back in winter, the roots safeguard the orange-like scent. A delightful plant to dry and mix into *pot pourri*; or add both fresh flowers and leaves to a summer salad or fruit cup to delight your friends.

Monarda didyma, bergamot or bee balm

Nepeta × *faassenii*, catmint, has aromatic leaves

***Nepeta*, catmints,** revel in the sunshine but appreciate a good loam about their roots, where the drainage must be good. They are perfect nosegay plants, with dainty blue flowers and grey-green strongly aromatic leaves. The whole plant is richly endowed with a minty aroma – even the stems – which is retained after flowering, though in a more musty range. A brush of the hand in passing is sufficient to persuade nepetas to give away their perfume.

***Nepeta cataria*, catnep,** is a native British plant of the hedgerows and good garden forms are available as seed from specialist seedsmen. A dainty plant with tiny white flowers, its mint-like leaves smell strongly of mint and pennyroyal. It is a rather straggly plant, and one to tuck among others. In May and June a few fresh or dried leaves infused constitute the catnep tea of former times.

Nepeta* × *faassenii is the catmint that forms soft decorative hummocks of deep lavender blue flower spikes and soft grey foliage in May and June (see p. 7 and above). Ensure good drainage for it, especially in winter, and choose a sunny position. Plant several to form a path edge or use them to decorate the side of steps. Invaluable used as a dry wall plant or in raised beds or on banks where it can be allowed to spread as ground cover. Propagate from cuttings or divide in autumn or spring. Use the flowers spikes and

A border of mixed *Nicotiana alata*, the tobacco plant

foliage in *pot pourri*. A fine tall-growing garden hybrid 'Six Hills' is not as readily available as it used to be.

Nicotiana Tobacco plants are often hispid (bristly) and tacky to the touch and produce a distinctive fragrance from their tubular flowers with flat starry mouths. Give them shelter from prevailing winds and plant in groups for the best effect. Put them near a patio or evening sitting-out corner because the rich fragrance is then at its best. They like a deep rich moist soil.

Nicotiana alata (often catalogued as N. *affinis*) is best treated as a half hardy annual and used in raised beds or tubs near the house. Although it is a bit languid during the day, the flowering tobacco comes to life in the evening, flinging wonderful scent to the wind. Modern hybrids stay awake much of the day, and their perfume is richer in the evening. The white and purple-flowered ones are strongest in fragrance, the green ones are scentless; the oranges and yellow are intermediate. With such improvements in flower power, some of the nicotianas are useful as bedding plants and for growing in tubs; they also flourish in pots to scent the conservatory. Many half hardy annuals sown in autumn will provide early spring flowers and nicotianas are no exception though they will be killed by frost.

Nicotiana sylvestris is one of the most dramatic tobacco plants, given a good sheltered border or grown in a tub or large pot. Treat it as a half hardy annual. It is a somewhat coarse plant but produces long trumpet-like white flowers in midsummer that burst and dangle from the top of the stem and give off a truly wonderful fragrance in the evening. Plant it in tubs on the patio or near a porch where the scent can be enjoyed through the open door – or for an almost overpowering effect bring the pots indoors in the evening.

Oenothera The evening primroses need a light well-drained soil and revel in full sunshine, but are equally satisfied with light shade. As the name suggests, most open their flowers only in the evening, when the rich fragrance is given off, and this is true of the biennials which tend to produce flowers so fleeting that they are replaced almost daily. But the perennial kinds retain their flowers longer, hold them open all day and in return forfeit fragrance.

Oenothera biennis, a naturalised plant in many areas, opens its sweetly fragrant flowers between six and seven o'clock in the evening to announce its presence to evening flying moths, and blooms over several weeks in late summer. During the day the petals cleave to one another and withhold scent. Put it in a sunny corner near the house windows to benefit fully from its late show, or scatter the seeds where the flowers can be shown off against a dark background in the fading evening light. It is a true biennial.

Oenothera caespitosa is a scented perennial known as tufted evening primrose, whose white flowers bloom over a very long period from May to August with a sharp magnolia-like fragrance in the evening. This is a plant for the front of the border or a raised bed where its rather straggly habit is best accommodated. The individual flowers are exquisite, with pleated white petals deepening to pale rose pink with age, almost without stems and lying close to the plant. Propagate by division in spring.

Phlox paniculata is the scented border phlox so frequently grown in both town and country gardens. It needs a fertile, moisture-retentive soil in dappled shade for real success. Totally intolerant of drought, the leaves of the border phlox wilt at the thought of a hot dry week. Strongly perfumed, pervading the air around, the scent varies from day to day. This is because the freshly opened flowers are far sweeter in fragrance than the mature and fading ones. Their true value as garden plants rests in their flowering period, July to

The border phlox flowers from July to September. This is *Phlox paniculata* 'Leo Schlageter'

September, when so many other summer-flowering plants are beginning to go over. In carnival colours from white to purple to geranium red, the flowers are held in domed heads that add strong colour emphasis in the flower border.

Primula This vast genus offers some superbly scented plants for a variety of cultural situations. Velvety flowers painted in the richest of jewel colours and sometimes a combination of colours, provide good garden effect as well as demanding individual admiration. Some that flower early are invaluable with the spring-flowering bulbs and for bedding schemes, and the polyanthus in particular, where the paler-coloured flowers usually carry the best perfume. For waterside and woodland planting, several scented species are ideal. All form basal rosettes of leaves and can be raised from seed and often divided successfully. Most are completely hardy, and a few, notably the scented *Primula malacoides* require greenhouse cultivation. *Primula viscosa*, a little rock garden plant, bears sweet-scented flowers but, unfortunately, evil-smelling sticky leaves.

Primula auricula has passed on its sweet honey-like perfume to all its derivatives over the centuries and the range of cultivated auriculas is now wide. Alpine auriculas, not seen as much as the border auriculas, are free from farina – the white meal that characteristically powders stem and calyces in the other sorts. Show auriculas are deliciously scented and are usually cultivated in pots and flowered under glass. Alpine auriculas may be used in the rock garden and flourish best tucked in behind a stone and facing north away from the direct sun. Give them some moisture-retentive material round their roots. Try a few plants at the front of a really shaded border among stones or in a made-up pocket of humus-rich soil or in a trough in the shade where the strong honeysuckle-like scent can be enjoyed.

Border auriculas bear the same rich scent too, but only share it when the atmosphere is really warm. Hardy and quite vigorous, they need a sunny spot in a soil that does not dry out in summer.

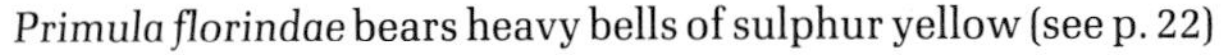

Primula florindae bears heavy bells of sulphur yellow (see p. 22)

Grey-green leaves with serrated edges are tempting to touch for their smoothness, and the flower are richly coloured. Many of the cultivars have been grown for many years – one of the fancier's flowers, and deservedly so.

***Primula florindae*, the giant or Himalayan cowslip,** (see p. 21), is one of the tallest-growing species and loves to be near water. It will grow, but not as luxuriantly, in drier situations. Heavy yellow bells of sulphur yellow powdered with meal, flower in June and July with a rich cowslip scent.

Primula helodoxa is sometimes called 'glory of the marsh', which offers a clue to its cultural requirements. Lovely rich butter yellow flowers in June and July are sweetly fragrant and borne in tiers up the stem. It makes a very handsome plant dusted in creamy farina. Give it a place where it gets sunshine for part of the day.

Primula sikkimensis has large wrinkled leaves and stout stems that carry a cluster of pendent funnel-shaped flowers in June. Sweetly scented like cowslips, they are a strong pale yellow, frosted with meal within, and really one of the most distinctive of the primulas. Choose the damper parts of the garden for best results and treat it as a biennial, ensuring shelter from both cold winds and prolonged sunshine.

***Primula veris*, native cowslip,** bears a distinctive sweet scent which, unlike many other flowers, disappears once the flower has been pollinated. Rich yellow flowers with a flared mouth and rounded petals are distinguished, upon examination, by a tiny red spot at the base of each petal. Creamy green, dry-looking calyces behind them seem somewhat inflated. Propagate from seed and grow it in grass, particularly on neutral or slightly chalky soils, in the way it used to be found in English meadows.

Primula vialii looks more like a polygonum than a primula; the small violet-blue bells are tightly packed among scarlet calyces in a poker-like head and strongly fragrant. The bright green leaves in a rosette are upstanding, lending a slightly military air to the plant. It may be rather short-lived but can be raised from seed.

Viola A genus which is at its best in cool moist conditions where the creeping tufted growth will flourish in moisture-retentive leaf-mould. Some violets have been grown for hundreds of years, and have been known equally long for their ravishing scent. They are good plants for really shaded areas or to grow in troughs or containers where the scent can often be appreciated. Many people say that violets are not always scented, because the perfume quickly anaesthetises the nasal nerves, rendering them inactive. So the first sniff is delicious and then the fragrance appears to fade.

***Viola cornuta*, the alpine or horned violet,** has a tufted habit and blooms among its heart-shaped leaves from May to August. The flowers are a pale lavender blue (there is also a white form) and especially scented at dusk. It is the parent of several cultivars such as 'Arkwright's Ruby', bright crimson with a dark blotch. Use this viola and its forms at the front of the border in dappled shade.

***Viola odorata*, the sweet violet,** whose flowers cooked or crystallised were used to provide a sweetening alternative to honey in days gone by. Their prim little faces appear from January to May depending upon the season and locality (sometimes even starting to flower as early as November). Propagate by dividing the stolons in April, or from seed, which germinates best in cool conditions. Some scented cultivars are more free-flowering than the type, for instance 'Nellie Britton' (sometimes listed as 'Haslemere'), rosy mauve, 'Baroness de Rothschild', lavender blue and 'Czar', deep violet.

Viola odorata, the sweet violet

Trees and Shrubs

More than any other feature in the garden, trees and shrubs need to be assessed carefully before planting, for they are the principal long-lived forms that will establish the garden pattern. Ultimate space required, season of interest, and suitability to the site are all factors to be taken into account; the type-casting has to be accurate. But when planning the scented garden there is the added consideration of fragrance – some scents are dominant, some short-lived, some strongest in the evening, others overpowering with fishy overtones – so select with care.

Small trees from which to choose, all with white flowers, might include the eucryphias, good for sheltered sites, *Halesia carolina*, the snowdrop tree, delicately perfumed, and the more robust *H. monticola* with a slightly stronger scent. Also good is the lovely rounded head of *Ligustrum lucidum*, the wax-leaf privet which flowers in October with strongly scented cream flowers. More white flowers are produced by *Stewartia serrata*, richly fragrant at midsummer and then putting on a foliage carnival show in autumn. Some would include the sorbuses, too fishy for many noses, but all with creamy white flowers, or the elders, hawthorns and, where wall space allows, *Drimys winteri* for its rich white waxy flowers in June and fragrant bark for the remainder of the year. Even more white flowers come with the magnolias – *M.* × *soulangeana* and its cultivars, all happy suburban dwellers flowering in March and April, and the lovely lantern flowered *M. wilsonii* with its never-to-be-forgotten lemon fragrance.

The laburnum (best as *L.* × *watereri* 'Vossii' which does not produce poisonous seeds) bears golden yellow flowers in long tassels in June and is justifiably called golden rain. For red flowers, the hawthorn might perhaps be chosen, whilst *Malus hupehensis* and *M. coronaria* have pink flowers in spring and leaves turning richly flamboyant in autumn.

Among the shrubs possibilities include olearia, osmanthus, rhododendrons, the latter requiring plenty of space and acid soil conditions, otherwise they seem so out of place, and sarcoccoca for

Magnolia × *soulangeana*, white with a faint flush of pink

first-class ground cover and *Hamamelis* or *Stachyurus* for winter charm. Consider stephanotis for the greenhouse or conservatory, with plumbago, the hoyas and *Mandevilla laxa* (*suaveolens*). Climbers are invaluable in the scented garden: roses, jasmine, lonicera, some clematis, trachelospermum, climbing clerodendrons and wisteria, especially *W. sinensis*, the best fragrance of all in the long mauve tassels (see p. 40).

***Acacia dealbata*, wattle or the mimosa of florists,** (see p. 1), is usually cultivated as a greenhouse or frost-free conservatory plant for very early spring flowering. In sheltered mild localities, such as the extreme south west of England, it is a tall shrub which may be grown outdoors against a sunny wall for protection. It loves the drought-like conditions at the base of a wall in a neutral to acid fertile soil. The buttercup yellow flowers are intoxicatingly fragrant of almond/violet and are carried in plumes of small fluffy balls. The foliage is feather-like, evergreen, and the growth somewhat lax. Propagate by seed, or from cuttings taken with a heel in summer.

Buddleja Buddlejas have a remarkable ability to attract butterflies, which, having once alighted, stay drunk with nectar. A lovely rich almond fragrance surrounds the flowers and drifts on the air also. Quick-growing, unfussy as to soil type, though perhaps happiest on the slightly more fertile soils, they flower when young; for quick results in the newly planted scented garden, plant one or two buddlejas. Propagate the kinds listed here by cuttings taken with a heel in late summer; *B. davidii* will also be successful from hardwood cuttings in autumn.

Buddleja alternifolia forms a large shrub eventually of weeping or arching habit, bedecked with mauve flowers in July and August, richly fragrant of heliotrope. Use it as a lawn specimen or high point in a small mixed border and watch the moths flutter about it in the evening.

***Buddleja davidii*, butterfly bush,** is probably – together with the richly scented elderberry – the most maltreated shrub of all, possibly because it is coarse-growing and quick to regenerate. Allow it plenty of room – its cane-like growth will peer over every fence – and choose a rubbly lime soil for best results. The buddleja will never fail, whatever the locality, to produce great arches of densely packed flowers in July and August which give off a rich musky honey-like fragrance. It is available in a range of flower colours through lavender and purple to rich burgundy, all of which are more fragrant than the white-flowered cultivars.

Buddleja fallowiana is a smaller and far more elegant shrub than *B. davidii* with similar flower and leaf form and a delicate fragrance. The stems and leaves are white and woolly, lending a frosted appearance and demanding some slight protection from winter dampness. Grow it in a sheltered position protected by a building or hedge where it can be seen against a darker background when in flower in July and early August.

***Chimonanthus praecox*, wintersweet,** has sulphur yellow flowers strung along rather whippy branches in February and March. They blow a very sweet fragrance on to cold, dry air. A shrub to delight the nose just as the days lengthen; later, when the leaves appear, they are impregnated with a restrained version of the sweet fragrance. It is a medium-sized shrub to plant where it will get plenty of summer sunshine to ripen the wood. Propagate from cuttings taken in summer and then wait, because it is reluctant to flower when young. Choose a sheltered spot in a well-drained friable soil, chalky if possible.

Choisya ternata Mexican orange blossom, so called because its lustrous evergreen leaves contain a rich spicy pungent orange scent released on crushing. The white flowers with a darker eye are gathered into clusters and have a far sweeter deep perfume like that of orange blossom. Summer is the main flowering time, but when it is happy, flowers will appear intermittently in flushes all year round. It is a medium to large shrub and a good hedge plant in the south of England, where it can be protected from wind chill, on a well-drained soil. Propagate in late summer by cuttings with a heel.

Cytisus The brooms have a reputation for being short-lived, but nevertheless their adaptability renders them immensely useful for the fragrant garden. They are sun worshippers, best on neutral fertile soil with good drainage, and flower when quite young with a wonderful outburst scenting the air around. Some smaller ones such as C. *purgans* with its upright habit are very sweetly scented and useful in tiny gardens or for cultivation in raised beds. In more extensive areas the arching branches of C. × *praecox*, the Warminster broom, will make fountains of sprays of sulphur yellow flowers early in summer – especially wonderful where several can

Cytisus battandieri flowers in July

be planted together as on a bank. Propagation is from cuttings (with a heel) after flowering, or from seed sown in spring.

Cytisus battandieri is one of the most handsome members of the broom family and clusters its flowers together into a long pineapple-shaped head, which is also strongly scented of pineapple. Set it in a corner or where it can gain shelter and encouragement from a house wall for its semi-evergreen foliage and soft woolly branches. It will soon reach to the top of a high wall. The flowers are bright yellow in erect heads in July and the silver branches are graceful and decorative for the remainder of the year.

Cytisus × spachianus is called the fragrant broom in spite of all its scented relatives. When allowed to grow naturally it is a tall spreading shrub, but often it is cultivated as a pot plant and marketed as '*Genista fragrans*'. Grown this way it can be relied upon to give a first-class performance in late winter and early spring. The scent trapped in a conservatory or warm room is ravishing. Keep the growth cut back constantly to retain the bright jade green foliage.

Daphne The veriest amateur gardener knows of the thrill of the rich spicy scent of *Daphne odora* during bitterly cold February and March days, and this is only one of the family that can provide fragrance some time from November to June. Some are difficult, reluctant to stay, others are like spoiled children – choosing to behave only when it suits them. But they are tolerant of most soils, provided drainage is good and there is an all-the-year-round supply of humus enrichment. Buy in pots or transplant rooted cuttings direct to the flowering site because most daphnes transplant badly. Propagation is from heel cuttings taken in summer. Try the tiny *D. alpina* in the rock garden or in troughs for its sweet-scented white flowers in May, or *D. retusa*, a miniature with evergreen leaves, for a pan in the greenhouse, or a trough on the patio. It has waxy rose-purple flowers in May and June.

Daphne blagayana bears a rather elusive scent, but its rich creamy flowers in March and April are a delight. It is rather temperamental but a challenging plant for a good leafmould-rich soil in very light shade. When the straggling branches are buried, branch tips emerge to form a circle of dancing flowers around the bush. A treasure where it is happy.

***Daphne* × *burkwoodii*,** (see p. 40), is a fairly quick-growing shrub of medium height smothered in pale pink deliciously scented flowers in May. 'Somerset' is a selected and perhaps better form to grow.

***Daphne cneorum*, garland daphne or garland flower,** (see p. 40), is a delightful little shrubby plant for the rock garden, where it loves to have its feet in the shade and its head in the sun. Tight clusters of red buds open to starry-mouthed pink flowers in May and June. Slightly waxy in texture, their scent is ravishing. Often it repeats the flowering performance in very late summer and, as the polished foliage is evergreen, the plant is an asset all year round.

***Daphne mezereum*,** best known for its highly polished red berries in summer, produces winter flowers in shades of mauvish pink to deep plum purple. In some areas they appear before Christmas, but generally the scent is associated with January and February, defying wind and rain with annual regularity. Select a wooded area for this winter treasure to trap the scent in the cold air.

Jasminum The jasmines are a versatile tribe of lax and climbing shrubs, all of which need constant tying in and frequent light pruning. The hardier ones tolerate town conditions and are not fussy as to soil provided that the drainage is good. Some, such as *J. polyanthum*, are best grown in a conservatory or porch where the scent can be retained and support provided for the pliant stems. Its long deep pink buds open to ravishingly scented blush white flowers which are produced very freely in early spring under glass, or a little later in really sheltered corners out of doors. For troughs or the rock garden try the miniature cushion forming *J. parkeri* with yellow richly sweet fragrant flowers in midsummer. Jasmines are propagated from stem cuttings taken with a heel after flowering.

***Jasminum officinale*, common jasmine,** is a vigorous scrambler that loves to entwine with some other plant. Plant it in a mixed border where it has some shelter and part of the day in the shade, and give it a 'wigwam' of supports and encourage it to cover that. Although it will look pretty bleak in the winter the white summer flowers will compensate. A rich fragrance will drift over the border and the plant flowers continuously from midsummer to early autumn. It will also scramble over an arch at a doorway, where its scent can be enjoyed in passing.

***Jasminum revolutum*,** a semi-evergreen for milder districts and town gardens, produces small, deep bright yellow flowers with rolled back petals from mid- to late summer. The gentle scent is one of the joys of the summer garden, and the plant does best on trellis or some support, or when it is allowed to grow through another plant. It does not object to being constantly restrained.

***Ligustrum ovalifolium*, common privet,** is in fact so common as a hedge plant in older suburban gardens that its value in the decorative garden tends to be overlooked. Its golden-leaved form, 'Aureum', when allowed to grow naturally, is a handsome upright large shrub. Buttercup yellow leaves with bright green suffused along the central vein look good all summer, and are evergreen or semi-evergreen in all but the coldest areas. Creamy white small terminal panicles of flower waft a heavy, almost sickly, rich scent across the garden in July. Privet is totally undemanding of soil provided it is not waterlogged; it accepts sun or shade, town and country, and can be relied upon to give of its best. Propagate from summer cuttings with a heel, or from hardwood cuttings, taken in autumn.

Lonicera The honeysuckles are among the most rewarding of

Jasminum polyanthum on a conservatory porch

scrambling scented plants. They are highly decorative and the climbing sorts always twine from east to west. Choose dappled shade or a place that has shade for part of the day on well-drained soil in which some moisture retentive material has been incorporated. Encourage them to enfold the pillars of a pergola or unfurl over an arch or tripod of supports. They are invaluable for disguising posts or old tree stumps. When cut back and allowed to flop about, the common *L. periclymenum* will form a large dense scented mound, rather difficult to disentangle for pruning, but nevertheless providing good scented ground cover. The perfume is richest and strongest in the twilight when the long-tongued moths that fertilise the flowers are on the wing. Berries follow most flowers and seed forms in good summers, so propagation can be from seed when ripe but more commonly from late summer or autumn cuttings.

Lonicera fragrantissima, a medium-sized shrub, not a climber, comes into flower for the new year and continues until spring in all

Above: Honeysuckle, *Lonicera periclymenum* 'Belgica'
Below: *Mahonia × media* 'Charity' flowers in winter

but the vilest of winters. The foliage is leathery, semi-evergreen and decorated with creamy flowers in pairs, held back to back. As with many winter-flowering plants. a site that catches the early morning sunshine should be avoided.

***Lonicera periclymenum*, common honeysuckle** of the hedgerows along with cultivated forms are winners in the garden and easy to manage. Long creamy white flowers in clusters are often flushed with purple-red on the outside, opening paler from June to September and even October in warm autumns. 'Belgica' with plum red flowers, and 'Serotina' with deep red in the flowers, are forms of *L. periclymenum* and are just as versatile and useful for their colour variation, retaining the powerful sweet fragrance.

Mahonia* x *media is the group name for hardy hybrids with stout stems from which a terminal cluster of flower stems radiate like spokes from an umbrella. They justly deserve a place near the house where the evergreen foliage and pale yellow flowers smelling of lily-of-the-valley can be enjoyed in winter. Good fragrant cultivars include 'Charity' and 'Winter Sun'. The mahonias like semi-shade and most good friable soil types, and are particularly useful for covering banks and as informal hedges. Propagation is from seed when ripe, layers in spring, or summer cuttings of stem tips.

***Myrtus communis*, common myrtle,** when grown out of doors in a sheltered locality produces an abundance of white starry flowers with fluffy centres in May and June (see p. 4). The whole plant is richly and delicately scented, and the foliage aromatic, but liable to damage from cold winds and frost. It is particularly good in seaside gardens, and a lovely scented plant for the conservatory or garden room when grown in a tub in potting compost. It is a plant that will thrive in town gardens too, and adapts to most soils including chalk. Propagate from cuttings taken with a heel in summer and give some heat to encourage rooting.

Philadelphus The mock oranges are a floriferous tribe of medium to large shrubs with a tendency to lax growth. The foliage is pale green and there is a general air of gentleness about them. The white flowers, abundantly borne, are powerfully scented of orange blossom and are so generous with it that the whole garden can be dominated by it being carried on the breeze. A place in full sunshine on dry soils suits them best and by careful selection of species and hybrids the period of flower can be extended considerably from May until July. Try the cultivars of *P.* × *lemoinei* like 'Avalanche' for their decorative flowers which literally weigh down the arching branches. Propagation is from hardwood cuttings of ripened wood in autumn or summer cuttings with a heel. Seed is produced, but does not breed true.

Philadelphus coronarius is the most fragrant-flowered species, and one of the most popularly grown, especially in its golden leaved form 'Aureus'. It is a good plant with which to lighten up a border, even a shaded one, and the leaves quieten down to a pale green-gold as summer progresses. The flowers are creamy white, weighing down the whole branch in June.

Rosa Fragrance is the universal appeal of roses. As climbers or shrubs there are numerous ways in which they can be incorporated into the decorative garden scheme. In general the climbers will enjoy more shade than the bush types and they all love a moisture-retaining soil containing a good supply of humus. More than any other plant roses display an enormous range of scent within the 'rose' spectrum – from fruity to tarry, to tea, to lemon, to spicy, while many more nuances are detectable. The species are the oldest roses in cultivation and are among the richest in scent. *Rosa centifolia*, the cabbage rose, with its full rounded flowers, is deeply rose-scented and its forms such as 'Fantin Latour' (see p. 36), pale pink, and 'Tour de Malakoff', magenta-mauve, retain the sweet full-bodied perfume. The damask rose, *R. damascena*, provides a gentle true rose perfume; try it in its forms 'Comte de Chambord', 'Marie Louise' and 'Celsiana', all pink. Some roses have not only fragrant flowers but also deliciously scented foliage. This is so in *R. rubiginosa*, the eglantine or sweet briar, a rose to plant so that the prevailing breeze will carry the aroma towards the house on a warm evening after rain – nothing could be more perfect. The arching branches are very prickly and bedecked with red hips in autumn. 'Lord Penzance', pollen yellow; 'Lady Penzance', copper gold; 'Amy Robsart', strong pink; and 'Meg Merrilees', crimson, are all favourite *rubiginosa* forms with delicately fragrant flowers.

Bourbon roses are free-flowering and some are fragrant over a long period from June to September. Some suggestions of varieties to plant are: 'La Reine Victoria', rose pink and true rose-scented; and 'Mme Isaac Periere', deep rose red and scented of raspberries. Cluster-flowered roses (floribundas) are sometimes devoid of scent but one or two may be included in the fragrant garden – 'Ma Perkins' for example, pink with a true rose perfume; 'Victoriana',

Rosa damascena 'Comte de Chambord'

tangerine/silver; and 'Golden Fleece', yellow and also smelling of raspberries.

The musk roses are vigorous shrubs which carry dense trusses of flower followed by innumerable glossy hips and are therefore good for autumn effect with the remarkably attractive foliage. Select 'Buff Beauty', apricot with an intense rose scent; 'Penelope', cream with a musky rose scent; and 'Felicia', pale silvery pink and spiced with an aromatic tang.

The large-flowered roses (hybrid teas) are legion, thoroughly deserving their popularity, but for fragrance try such as 'Doris Tysterman', tangerine; 'Fragrant Cloud', coral-scarlet; 'Papa Meilland', crimson; 'Super Star', vermilion; 'Troika', amber; 'Whisky Mac', golden orange; and 'Blue Moon', lavender (not to everyone's taste for colour, but lemon-scented). A few miniature roses have fragrant flowers and are useful for the rock garden edge, troughs, walls or in pots; 'Sweet Fairy', pale mauve; 'Yellow Doll', pale yellow, and the white 'Twinkles' are all scented.

Of sterling value in the scented garden are the climbing and rambler roses, for they lift the scent up to the air and they combine well with all kinds of other garden plants, which bush roses do not do as happily. Many are especially useful when grown on walls, where the scent is enormously enhanced in the evening by the reflected heat from the wall which keeps the temperatures up to coax even more perfume out of them. Fragrance rather than the classification of type is the concern here for there are numerous hybrids and cultivars, sports and species of climbing habit. Some are good in one situation, others in another and they are used to clothe fences, walls, arches, pergolas and posts or to fling up into an old tree, to festoon an arbour, camouflage a shed or decorate a balustrade. One or two to look out for are *Rosa filipes* 'Kiftsgate', vigorous, white-flowered and smelling of incense; *R. laevigata*, the Cherokee rose, deep cream and spicy; 'New Dawn', shell pink and enticingly fruity; 'Paul Lédé', apricot-buff and smelling like tea; 'Zéphirine Drouhin', soft rose pink and sweetly scented; 'Goldfinch', yellow fading to cream and strongly scented; and 'Veilchenblau', lavender-blue-grey-mauve and scented of oranges (plant it where it can be shown off against a light background).

It is nearly impossible to select 'the best' for the fragrant garden but for special interest the following are recommended (see also the Wisley Handbook, *Roses*).

Above: *Rosa filipes* 'Kiftsgate', a vigorous climber
Below: *Rosa centifolia* 'Fantin Latour' (see p. 34)

***Rosa centifolia* 'Muscosa', the moss rose,** is a small shrub in which the scent glands can be seen on the backs of the petals and as a soft bristly green growth on stems and calyces. Even when the flowers are in bud the moss-like growth releases a resinous scent which is retained on the skin. There is a slightly oily feel to the flower head and the fragrance is full and richly rose-scented. A good plant for the small garden, and nice in the forms 'Mousseline', for its intense scent, or 'William Lobb', with a truly captivating rose perfume.

***Rosa banksiae* 'Lutea', yellow Banksian rose,** outstrips many others in its race up the house wall, and produces a myriad of tiny yellow double flowers quite early in the summer. Grow it where space allows, for its delightful violet scent.

***Rosa gallica* var. *officinalis*, the rose of Provins or apothecaries' rose,** although prone to suckering, is one to introduce for fragrance, especially where space allows this small spreading bush to be planted *en masse* – or to make an informal hedge. It is best of all for drying the petals to incorporate into *pot pourri* for they retain all the fullness of their scent. Deep red semi-double flowers appear in June, on low bushes that can be regularly pruned hard and will regenerate with vigour. For a true rose perfume with all its depth and enchantment, 'Belle de Crécy' is the form to grow, with its magenta pink flowers fading to crimson and parma violet.

Syringa The common lilac has been grown in our gardens since the sixteenth century and forms a tall upright deciduous shrub well known for its strong scent. It was introduced from the mountainous regions of eastern Europe and the Middle East, along with the highly scented *Philadelphus* (see p. 33). In common parlance the two have been confused ever since; the one, *Philadelphus*, being known by the name *Syringa* of the other. The lilacs are unselective of soil, but show perhaps a little preference for chalk, and grow equally well in both town and country. Some of the species are highly fragrant and the rose-pink flowered hybrid *S.* × *josiflexa* (*S. josikaea* × *S. reflexa*), raised in Canada, particularly so. The Rouen lilac, *S.* × *chinensis*, with drooping flower panicles of soft lavender is heavily fragrant and has been known in cultivation for about two hundred years. Propagation is from summer cuttings taken with a heel, or hardwood cuttings in autumn.

***Syringa* × *persica*, Persian lilac or blue Persian jasmine,** claims attention for the smaller garden, as it forms a small rounded

Rosa gallica var. *officinalis*, the apothecaries' rose

graceful shrub decked with typical lilac-shade fragrant flowers in dainty panicles in May. Best protected from cold winds, it is otherwise a gem. The scent is light and distinctly spicy.

***Syringa vulgaris*, common lilac,** provides a host of cultivars with heavy panicles of flowers in May varying in colour from white through mauve to deep plum red. They are richly but variously scented, wonderful after rain, and good summer screening shrubs especially useful in town and suburban gardens. Old flower heads ought to be removed but rarely are, and pruning is neglected also, but still the shrubs flower with unfailing regularity. Good heart-shaped clean green leaves make a dense foil for the flowers. Cultivars are numerous, but some well-tried ones are: 'Congo', deep velvety purple flowers in rather dumpy heads; 'Mme Lemoine', top-of-the-milk cream in bud and pure white double flowers; 'Charles Joly', another with double flowers in plum red and flowering rather late; and 'Ambassadeur', blue-mauve flowers with a white eye.

Viburnum Many viburnums bear flowers strongly scented of honey and the deciduous ones follow with a good autumn display of shining berries and startling foliage. Viburnums dislike a peaty soil, but are happy in most situations and tolerate seaside gardens and town gardens alike; *V.* × *burkwoodii*, especially, seems to thrive among town buildings. Propagation is from cuttings with a heel, taken after flowering for the deciduous ones, and in early summer for the evergreen ones.

***Viburnum* × *burkwoodii*,** an evergreen, is one of the most useful medium-sized shrubs for winter flowers. Encrusted flower heads of deep pink buds appear about the new year, open to white with a really sweet scent, and go on into April and May.

Viburnum carlesii forms a small rounded deciduous bush with often particularly good autumn colour. It is deservedly popular for its ease of cultivation. The flowers are sweetly daphne-like or spicy in scent, and bloom in April and May. Buds are rosy pink, and open slowly to give pure white flower heads.

***Viburnum farreri* (*V. fragrans*),** which Reginald Farrer called 'the most glorious of shrubs', is another winter-flowering plant. The foliage is attractively bronzed when young and the pink bud clusters turn to white on opening. They often manage to do this before Christmas and go on for some time during the darkest days of the winter. They are richly scented with a spicy overtone.

Wisteria sinensis forms a rather gawky shrub, seen at its best when supported by a wall or pergola or allowed to grow over a shed or summerhouse. Then the long tassel-like mauve flowers can hang down among the filmy decorative foliage. Plant it somewhere where its nakedness is masked in winter to spare its awkwardness, and provided there is some moisture-retentive material deep round the roots it will outlive two or three generations. The flowers in May are fragrant of vanilla when the entire plant assumes a likeness to a Chinese painting on silk. Propagation is from summer cuttings with a heel and some heat.

Above: *Wisteria sinensis* flowers in May
Below left: *Daphne cneorum* flowers in late spring (see p. 29)
Below right: *Daphne* × *burkwoodii* 'Somerset' (see p. 29)

Bulbs and Corms

One of the very best investments for the garden is bulbous plants, provided that good-quality material is selected. Once planted in a suitable position, most of them look after themselves, giving pleasure for years, and even increasing in number. The most widely cultivated are those suitable for mass planting. Early spring holds much of the garden excitement for scent when bulbs have been planted of narcissus, crocus, muscari, snowdrops and iris. However, beginners often fail to realise that autumn-flowering species of snowdrops, crocus and cyclamen are also available. Careful perusal of catalogues, or advice from specialist firms will suggest further ideas. (See the Wisley Handbook, *Growing Dwarf Bulbs*.)

Out of doors the scent of plants that bloom in the cold dim days of winter is often lost on the wind, or can really only be enjoyed by close inspection. So plant *Crocus chrysanthus* 'Snow Bunting' and miniature narcissus, such as *Narcissus* 'Silver Chimes', and the honey-scented snowdrops like *Galanthus* 'Magnet' or 'Straffan' in troughs near an entrance to the house or garage where they may be enjoyed. Often the mad March winds coincide with the flowering of some of the scented plants such as narcissus, muscari, *Iris reticulata* and *I. histrioides*, but where space allows, if these are planted in drifts, the somewhat elusive scent will be sensed occasionally. All can be cultivated in pots, troughs, or window boxes to give their special delights.

As for summer-flowering bulbs, lilies probably claim the accolade for scent, but galtonia, summer irises such as *Iris florentina* and *I. pallida* and pancratium are all worth a place in even the smallest garden.

One great joy of bulbs is that they can be grown in pots. The right growing conditions can be given, and they will reward with unblemished highly fragrant flowers in the conservatory, porch, living room or on a loggia or patio in summer. They can be transported from place to place to give the right colour emphasis in a summer flower border – lilies can be used in this way – and they can even be given or lent to friends, hospital patients or used for church decoration.

Crocus ancyrensis provides a cheerful start to the year, but it needs sunshine to persuade the flowers to open

Galtonia candicans flowers in July and August

Crocus Crocuses are undoubtedly at their best when naturalised beneath trees or among shrubs, but a few at the edge of the lawn where they can be enjoyed during the short winter days, when the sun coaxes them to open, are a delight. Unselective of soil, they are so defenceless against mice and birds!

Crocus ancyrensis is one of the first scented crocuses to flower, probably in time for the new year, and it will startle with its yellow spears that burst into glowing tangerine in the slightest ray of warmth from the sun. Put it in a spot that catches the early afternoon sun, for this one reason.

Crocus chrysanthus usually flowers by mid-February except in particularly harsh winters, and is the parent of a wide range of seedlings such as 'Blue Giant' and the purple and white 'Ladykiller'. But the best for scent is the pure white 'Snow Bunting', feathered and streaked with purple.

Galanthus Numerous garden forms, quite indistinguishable to all but the practised eye, have a refreshing spring-like scent reminiscent of primroses. Although by selecting the forms and species, snowdrops can be in flower from October to April or even May, they are beloved for their toughness and purity during the worst of the wintry weather.

They cannot be described as richly scented, but that spring freshness is a mossy scent on the cold air, and one or two are slightly violet-scented. They are harbingers of spring that hate the soil to be fed; if you wish to move them, they are best divided just as the flowers are fading.

Galanthus nivalis is probably a native British plant, with narrow strap-shaped leaves and globular white flowers that dangle by the merest thread from the top of their individual stem. Varieties such as 'Magnet' and 'Straffan' are distinctly honey-scented.

Galtonia candicans produces white trumpet-like bells tinged with green all around a 2½ foot (75 cm) stem in July and August. It is perhaps seen at its best in a sunny sheltered site in humus-rich well-drained soil where it is allowed to develop good clumps. These may need covering in the winter, as it is a South African plant. Its real value as a scented plant is after the stems have been cut and arranged in water, for then the fragrance strengthens. A lovely plant for church flowers for an August wedding.

Hyacinthus orientalis forms are nowadays limited to the bold compact clusters of flowers held on fat straight stems, and displaying no resemblance to the species. The common hyacinth is seen at its best when forced in pots or bowls. Because the double-flowered forms do not force as well, they are not offered for sale as readily. When grown in bowls a good fibrous bulb compost should be used and the bulbs close planted in September or October. A deep rich fertile soil and a position sheltered from the wind are needed for success out of doors, when bulbs should be planted almost 5 inches (12.5 cm) deep in September or October. They need to be lifted each year after flowering, unlike most bulbs. The scent is penetrating, heavy and sweet. Some popular varieties from which to choose are 'L'Innocence', white; 'Myosotis', blue and particularly richly scented; 'Lady Derby', pink; 'Yellow Hammer', yellow; and 'Cherry Blossom', red.

Iris Most irises appreciate open sunny positions, where the drainage is good and there is at least a smattering of lime in the soil. Some are rather reluctant to flower, but those listed here should not present difficulties for the average gardener, especially when one's efforts are reinforced by good summers to 'bake' the iris clumps after flowering. The great rainbow range of cultivars of *Iris germanica*, or bearded iris, have a rich fruity scent which thickens and changes once the stems are cut and arranged indoors.

Iris histrioides defies the elements to bloom from mid-January onward, with gorgeous deep blue spiky flowers marked with gold that nestle close to the ground. For this reason alone, it is best to cultivate this iris in pots and bring them into the garden room or porch where the cool sweet perfume can develop and be enjoyed from time to time. The form 'Major' is usually grown, with ultramarine blue flowers and black and golden markings.

Iris florentina provides the scented orris root of commerce, and is a plant of ancient cultivation. The roots possess the fragrance of violets, which strengthens as they dry after lifting. The flowers in May and June are white tinged with lavender with a golden yellow beard on the falls, and sword-like leaves. A good mid-to-front of the border plant which will make a fine clump if left undisturbed.

Iris pallida is known for its sweetly scented flowers although, because it is frequently seen in cultivation in its variegated leaf form, it is probably considered first for its decorative foliage. The

Iris reticulata flowers in winter

two forms with coloured leaves are 'Aurea Variegata' and 'Argentea Variegata', both with pale blue flowers but with a strong vanilla scent.

Iris reticulata, a winter-flowering species, is totally reliable, producing its curiously upstanding blue-mauve flowers in the severest of winters. It is a good plant for the rock garden, trough or for bowls in the conservatory; grown in crocus bowls (with planting holes all round) and brought indoors, this iris makes a delightful violet-scented table decoration. Choose a really well-drained spot for outdoor cultivation and grow 'Cantab', with pale blue flowers, or 'J. S. Dijt', plum purple.

Iris unguicularis (see back cover) is a winter favourite which unfortunately hides its pale lavender-blue flowers among its papery leaves. If you gather the blooms and bring them into a warm room, they will give off their violet scent and bring spring freshness to a January day. Plant at the foot of a wall, especially where it can be

Lilium regale has very fragrant white flowers (see p. 51)

sun-baked during the summer, and forget it for years. That is the best encouragement to flowering.

Lily The amateur gardener often regards the cultivation of lilies as something of a challenge, but by careful selection, i.e. suitability to site, or cultivation in pots, good results can be assured. Not all lilies are scented, but those that are usually emit a strong rich scent, even too strong for some tastes. When grown in pots or tubs the bulbs need to be planted deeply if they are stem-rooting lilies like *Lilium auratum*, and just to nose depth otherwise. Good drainage is essential and any good proprietary compost will suit them. Leave enough space at the top of the container to allow for top dressing with additional compost later. Buy from reliable sources, preferably in October, and the fleshy-scaled bulbs ought to be plump and firm. The great advantage of growing lilies in pots is that they can be moved about the garden and patio when in flower to gain the best effect and full enjoyment of their luscious flowers.

Lilium auratum lives up to its popular name of golden-rayed lily by producing abundant very large white flowers broadly decorated with a band of waxy gold along each petal. It is normally late-flowering, so is best grown in a pot, to get an early start to growth. Several hybrids are available, all richly endowed with spicy scent, and they flower over a considerable period in the late summer. Good on acid soils where sharp drainage can be provided together with shelter from prevailing winds.

***Lilium candidum*, madonna lily,** is best when grown in borders where it can be left undisturbed for a number of years. Find it a spot where it can stand in the shade and hold its head in the sun. The glistening pure white flowers, assembled in a cluster at the top of the stem, are decorated with bright golden anthers and an almost overpowering honey scent – too strong for some people. This is perpetuated in its descendants, the Cascade Hybrids. After flowering the bulbs are dormant for only a short period, and need to be planted or transplanted in August.

Lilium henryi is one of the easiest lilies to grow, especially on chalky soil. It gives a display of deep apricot orange flowers spotted prominently with brown. Sometimes there are as many as 40 or 50

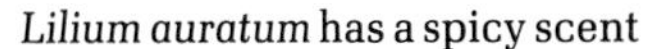
Lilium auratum has a spicy scent

Lilium henryi is an easy lily to grow

flowers to a stem in July or August. The sweet scent is particularly lingering and has found its way into the range of Aurelian Hybrids which have *L.henryi* in their parentage.

***Lilium longiflorum*, the Easter lily** of florists' shops, is a good cool greenhouse plant where it can be gently forced, and never fails to bloom in March and April. But out of doors, in sheltered areas the flowers come in high summer when the evenings are long and the jasmine scent can be enjoyed. A good 'special event plant' for a border near house or patio; but doubtfully hardy. Best in the variety *takesima* with somewhat grander pure white trumpets up to 6 inches (15 cm) long. A number of named varieties are available, the best perhaps 'Croft' and 'White Queen'.

Lilium parryi revels in dappled shade and is good for an open woodland planting. Its clear yellow funnel-shaped flowers stand up well against darker backgrounds. It is a lily of exceptional grace, flowering in May and June.

Lilium regale is one of the easiest of lilies to grow, and is justifiably popular. Lime-tolerant, it likes a sunny spot where its roots are protected by its neighbours from winter frosts, which suggests a shrub border as an ideal site. The white flowers are strongly fragrant, funnel-shaped and flushed brown and maroon pink on the outside and deep golden yellow at the throat within.

Muscari Leave muscari bulbs undisturbed for four or five years so that the colonies will build up, then divide them. They are most attractive when planted together with other bulbs and naturalised under deciduous trees or shrubs, for they love every ray of winter sunshine. These plants are tolerant of lime, but happy on any soil but the coldest of clay. Their grass-like leaves usually peep through in autumn.

Muscari armeniacum is a really easy plant, thriving equally well in town and country gardens. In April and early May the deep blue flowers edged with white are knotted about the head of the stem. The form 'Cantab' is popularly available, with slightly paler blue flowers and the same sweet fragrance. Planted in a pocket on the rock garden or as a ribbon edge to a border or path, it cannot fail to attract attention. It is not too demanding as to sunlight; semi-shade appears to satisfy its needs.

Muscari moschatum is the most sweetly perfumed of the muscaris, with a musk-like overtone, as its specific name implies. Perfectly hardy in a sunny position, it has rather dusky purple green flowers (that are not to everyone's liking!) which turn to a golden olive green lower down the stem on maturity. It blooms in April, with leaves rather broader than those described as grass-like.

Narcissus The name narcissus relates to *narc* – a dullness of sense – from whence is derived the word 'narcotic'. This is because in some forms the perfume of narcissus induces headaches and giddiness when deeply inhaled. Ordinary daffodils are undoubtedly the most popular bulbs for naturalising in grass, among shrubs or in woodland areas, and yet they are equally effective when cultivated in pots, tubs, window boxes or even old wheelbarrows! Symbolic of the English spring, some of the flowers are strongly scented, others not. Most seem to prefer the meadow/woodland type of soil and many of the small ones are happy where drainage is better, for example, in the rock garden or in troughs.

Garden cultivars of narcissus are divided into twelve divisions, dependent upon the type of flower. All have a fresh spring earthiness about them, but some have a deliciously rich scent, sometimes quite overpowering. Bulb catalogues are a good guide to the innumerable cultivars available and usually record the scent. All narcissus bulbs need to be planted early; August is not too soon, as the new roots start to grow again in the late summer. (See the Wisley Handbook, *Daffodils*.)

Jonquillas, suited to cultivation in sheltered areas out of doors, or in pots in the conservatory or indoors, are perhaps the strongest of all in their perfume. The fragrance is quite distinctive, rich and spicy, and the golden yellow flowers with short cups are carried in little clusters at the head of the stem.

Narcissus juncifolius in the same group is a delightful miniature species, with dark rush-like leaves and tiny yellow flowers in clusters in March and April. It is ideal for cultivating in pans in the cold greenhouse or in a trough by a sheltered patio or a scree bed or pocket in the rock garden.

N. poeticus are the latest flowering of all the spring-flowering narcissi and sometimes get overlooked for this reason, coming during the plenteousness of the fragrant plants of May. Plant the bulbs in groups and leave them undisturbed and their chalk-white petals and crushed flat cups of pure gold, tangerine or yellow will provide gorgeous cut flowers with good stems for many years.

Tazettas are sweet-scented, bunch-flowered 'narcissi' which in the very mildest areas can be persuaded into flower out of doors for the new year. The popular 'Paperwhite', known to florists for the Christmas market, can be cultivated in bowls, or even in clean water, indoors, in a bowl of pebbles to anchor the roots. The scent will fill the room. 'Geranium', 'Soleil d'Or', and 'Cragford', with frilly cups, can be grown out of doors in mild places, and are excellent for forcing.

Pancratium illyricum is an unusual summer-flowering bulb, and one for those amateur gardeners who wish to try something a little different and challenging – best perhaps in a cold frame, or in a made-up bed in a sheltered spot in towns and other mild areas. Star-like fragrant white flowers, with tufted stamens and resembling daffodils in shape, appear in May and June. In autumn plant the large pear-shaped bulbs deeply for winter protection.

Aromatic Plants

Aromatic plants are the custodians of the richest and probably the most complex of essential oils, locking them away within the plant itself. Leaves often need to be crushed or rubbed to persuade the plant to release them to the nose. For the most part aromatic fragrances are subtle, less flowery in quality than flower scents, and are usually only discernible during the summer months even in some evergreen plants.

Herbs fall into this category, cultivated for centuries for their essential oils as flavouring and perfume, and while they may not be among the stars of the garden, they provide a valuable back-up chorus line, rich in fragrance. Their close association with man is recorded in the general use of their common English names – their scientific names seem somewhat unfamiliar: *Balsamita*, costmary, alecost or camphor plant, with minty-camphor-like fragrance and smooth leaves; *Myrrhis*, sweet cicely; *Levisticum*, lovage, with its earthy celery-like flavour and smell; and *Ruta*, rue, a good border plant with blue-green foliage, bitter in the extreme, but bearing yellow flowers that try to compensate by pretending to be cowslip in their scent. Other aromatic plants like the *Cistus* or rock rose produce their scent from the oily glandular hairs of the leaves, and some of them are even called gum cistuses. Conifers also are resinous, displaying a range of scent and always enhanced by a warm moist day.

Some green aromatics only really assume their real fragrance upon drying; *Galium odoratum*, woodruff, is one of these. Their place in the scented garden is assured, however, because they are lovely garden plants in themselves. And what could be more rewarding to the gardener, during a summer evening stroll round the garden, than to find a plant which, when touched in recognition, gives its scent in return?

***Aloysia triphylla* (*A. citriodora, Lippia citriodora*),** lemon verbena, is claimed to have the leaves with the strongest aroma. It survives out of doors tucked into a corner or sheltered border where there is a dry root run, in the milder localities, otherwise it is a good conservatory plant to grow in a container. Long narrow shining green leaves are highly fragrant of lemon, which they surrender when broken or rubbed. Propagate by cuttings in summer.

Artemisia Artemisias are perennials grown for their decorative aromatic foliage, and the scent is generally refreshing and sweet. Use them in the flower border to cool down the colour effects, or grow them beside a frequently used path where they can be handled often to release the fragrance. They all appreciate a friable soil and sunshine. Culinary French tarragon is a form of *A. dracunculus* which lacks the sweetness of most artemisias.

***Artemisia abrotanum*, southernwood or old man,** is a shrubby cottage garden plant with silky thread-like foliage with a crisp sweet camphor-like aroma. It is horribly gnarled in winter, so tuck it among better looking companions and cut it back in the spring to encourage new growth.

***Artemisia absinthium*, wormwood,** is a delightful plant forming a rounded medium-sized bush in a good moisture-retentive soil with its head in the sun. Silver green-grey leaves, with a tinge of apple green overall, shaped like little hands are sweetly fragrant with camphor overtones. Decked with dew or raindrops the whole plant is magic and the scent will linger about it. Cut it back judiciously to keep the fresh foliage going for the best scent. Add a few leaves to *pot pourri*.

***Artemisia chamaemelifolia*, lady's maid,** is a gentle plant, with green finely cut foliage that holds a sharp sweet camphor-like aroma when it is bruised. A good edging plant, because it tolerates clipping, it is also effective when used in block planting.

***Artemisia pontica*, Roman wormwood,** has a very clean camphor-mint aroma held in delicate foliage. The leaves are feathery and thread-like on twiggy upstanding stems. Good for ground cover in both sun and shade and particularly useful for awkward banks that can be given over to the one plant. Its rhizomes run about just below the surface so keep it well away from border favourites.

Artemisia tridentata has small three-pointed dull grey wedge-shaped leaves, which crowd together in clusters. It has a truly sweet aroma, outstanding even among the artemisias, and gives it up so easily on the surrounding air, but it is invasive.

***Chamaemelum nobile*, Roman chamomile,** often catalogued as *Anthemis nobilis*, is a useful plant because its tough jointed stems grow prostrate and so the plants are mat-forming. A perennial of

Artemisia abrotanum, southernwood or old man

ancient cultivation, it is still popularly used for making chamomile tea from the aromatic flowers. The thread-like leaves are also rich in aromatic oils. Its somewhat ragged appearance is compensated by its delicious apple fragrance and it tolerates being walked on so is invaluable in making scented paths and lawns. Grow the non-flowering kind called 'Treneague' for these purposes, or to clothe a bank behind a seat. Rooted stems need to be planted in friable stone-free and well-drained soil in April about 4 to 6 inches (10–15 cm) apart to form a 'mat'.

***Foeniculum vulgare*, fennel,** is a superbly graceful plant with finely cut foliage in soft plumes. When bruised these leaves give off a strong anise-like smell. It is a tall plant with an unusual soft feathery shape and is especially decorative in the form entirely suffused with a rich black purple bronze. The flowers are yellow in dainty flat heads in June and July, but to keep the aroma sweet and also to avoid excessive self-seeding, cut away the flowering stems when the buds form. Sow seed in spring or divide established roots in September and give fennel a sunny spot in a well-drained soil. It is usually a kitchen or herb garden plant, but of considerable worth in the flower border for both its foliage colour and form.

Helichrysum serotinum is redolent of curry, hence its popular name of curry plant. It is a perennial with a woody base and likes a position in the front of the border or across a bank where it can enjoy full sunshine. In the decorative border its needle-like silver foliage has a role to play, especially after rain or dew when it sparkles like tinsel. Dingy yellow flowers on rather straggly stems in July and August do nothing to endear it. Its curry supper smell travels quite a distance when the atmosphere is warm.

***Hyssopus officinalis*, hyssop,** is an aromatic shrubby little perennial which is evergreen in the milder localities and in the south of the country. The leaves are tiny, dark green and lock away its spicy fragrance until the warmth of summer. The aroma is at its richest as the tiny flowers appear in June. It is useful as a front of the border plant in a sunny spot and tolerant of dry conditions, but invaluable as a container plant throughout the year. The tiny snap-dragon-like flowers are fleeting and sparse, usually blue but there are pink- and white-flowered forms. It is a lovely old-world plant to enjoy for its fragrance.

Lavandula stoechas, French lavender, flowers in summer

Lavandula The lavenders are valuable garden plants cultivated since ancient times for their unique refreshing fragrance which permeates both leaves and flower bracts. They prefer good drainage and full sunshine away from frost pockets but even so are short-lived. Chalky soils seem to encourage a richness of perfume. Propagation is from cuttings taken in early summer and it is a good idea to maintain a supply of rooted cuttings to replace any losses in winter and to use in containers such as tubs and window boxes.

Lavandula angustifolia (syn. ***L. spica, L. officinalis***), **old English lavender,** is instantly recognised for its mound of grey foliage pricked all over in July and August by mauve flower heads which resemble pins in the pincushion. A well grown established plant is richer in scent than a young one and, as with most other aromatic plants, the essential oils are richest just before the flowers mature. This is the time to gather flowers for *pot pourri* and other fragrant conceits. Various forms of lavender are a variation on a theme: 'Grappenhall', with large pale mauve flowers; 'London Pink', pale pink; 'Munstead' (sometimes marketed as 'Munstead Dwarf'), good purple flowers on a compact plant; 'Hidcote', dark purple; and 'Twickel Purple', purple and the leaves set at right angles to the stem, with perhaps the best of all lavender perfumes.

***Lavandula stoechas*, French lavender,** is a smaller less splendid plant with green-grey leaves and 'hat pin' flower heads with prominent persistent bracts. The scent is far more oily and herby than that of *L. angustifolia*, which reminds us that lavender is an ancient antiseptic. It demands a really sheltered spot and is a truly short-lived perennial. Grow it in a container, and give it some winter shelter to help it through the worst winter weather. Propagation is from summer cuttings.

***Melissa officinalis*, lemon balm,** (see p. 59), is a true English cottage garden plant, important for the bee keeper and sometimes known as bee balm. In Tudor times it was grown as a strewing herb. It is one of the aromatic plants that retains its rich lemony fragrance after cutting, and dries well. Good rounded bushes solid in appearance are formed, so that melissa looks well at either side of a gateway or as a 'corner stone' of flower borders. The leaves are tough, nettle-like in shape and at their most decorative in the variegated form, 'Variegata', when they are painted with a clear yellow. The form 'Aurea' has an even more golden leaf area. Both are excellent for shade as they do not lose their colouring, although they much prefer a sunny spot with a good moisture retentive soil. If they dry out on light soils or during particularly dry summers, the aroma will deteriorate and take on a distinctly musty and stale overtone. Try growing it in tubs where the soil condition can be controlled. Little white flowers, rather detracting from the plant, appear in June and July. Propagation is best from cuttings of new growth taken in May as seed is slow to germinate. Cut back growth at the end of the summer to encourage a fresh crisply aromatic renewal the following spring.

Mentha The range of aromas is wide in the various mint plants and is matched only by their numerous English names. Most are outrageously invasive and unselective of soil, although they do need to have some moisture-retentive element to produce a clean fresh aroma. Late in the season and in dry summers their scent becomes very musty. All parts of the plant are impregnated by the aroma throughout the year. Propagation is usually very easily effected by breaking away a shoot or two with the creeping stem and a few roots. All the mints have a slightly military air and despotic intentions, so confine the roots in some way – bricks or slates pushed into the ground palisade-fashion are effective. The most refreshing clean scent of all is in *M. spicata*, spearmint, of the kitchen garden and used for mint sauce. None have remarkable flowers; they all form spikes of mauve or pink blooms and it is advisable to remove flowering stems to preserve the aroma at its best.

***Mentha* × *gentilis* 'Variegata', ginger mint or Scotch mint,** marches about but it is useful for its lettuce-green leaves with golden yellow suffused about the veins. They are warmly pungent of ginger when rubbed. Put it at the front of the border, where it can be restrained in passing.

***Mentha* × *piperita*, peppermint** and its forms, variously comprise eau de cologne mint, orange mint and bergamot mint, and herein lies the key to the fact that the composition of the essential oils varies imperceptibly, suggesting one scent today and another tomorrow. A plant grown in one situation may have a completely different scent to one grown elsewhere, but is always an interesting nuance of aroma. Often the leaves are rather heart-shaped and swirl around the square stem. *M.* × *piperita citrata* is highly fragrant and refreshing and has the reputation of enhancing the fragrance of its

Melissa officinalis 'Variegata', lemon balm (see p. 57), has lemon-scented leaves and is a vigorous herbaceous perennial

neighbours. The lovely jade green stems of *M.* x *officinalis*, white peppermint, add a decorative contrast to any collection of mints.

***Mentha pulegium*, pennyroyal,** makes a tiny-leaved mat-forming plant, useful to put pools of growth into a scented path, or to make a patch of mint-scented lawn. Pungent and peppery in aroma, it is good among paving or on a patio where it can be trodden upon occasionally to release the scent. At midsummer the upright form provides upstanding pale mauve flower spikes, so leave this mint to flower and enjoy its alternative performance.

***Mentha suaveolens*, apple mint,** is grown for its rounded hairy leaves with pretty edges and creamy white variegation. It is a most satisfactory plant at the edge of a terrace where the paving confines its activities and allows it the moist soil it needs. It is better in some shade despite the variegation, and is constantly sweet and fruity in aroma.

***Mentha* x *villosa*,** and best in its form *alopecuroides*, is Bowles' mint or French mint. Tall-growing and vigorous with soft downy wrinkled leaves smelling deliciously clean of fresh sharp mint with fruity overtones, it is good in a border where its wanderings can be tolerated.

Origanum Three kinds of marjoram are commonly cultivated as herbs and all are excellent aromatic garden plants. Use them as border or path edging or grow them in raised beds and dry walls near the patio, or cultivate them in window boxes where the leaves can be rubbed occasionally to release their scent. Let them relax somewhere in full sunshine, and ensure good drainage especially in winter. Propagation is by seed, cuttings or division of established plants.

***Origanum majorana*, sweet or knotted marjoram,** is usually cultivated as an annual and has brittle stems and soft grey, rather downy foliage. Dusky in general appearance, the plant produces pale lilac pink and white flowers in June and July held in little blob-like heads among the knotted clusters of leaves. Its fragrance is sweet and quite the best as a flavouring in cooking.

***Origanum onites*, pot marjoram,** is somewhat rougher and more pungent in aroma but highly aromatic. It is a popular plant and the aroma persists to a certain extent even through the winter. Cut back the growth in spring to encourage fresh shoots. The flowers are pink or white, from June to August, and the leaves rounded; the whole plant forms good little hummocks. Marjoram is especially useful at the path edge where it can sprawl about.

***Origanum vulgare*, wild or common marjoram,** sometimes confusingly called oregano, is easily recognised because it is altogether pinker in appearance than the other marjorams. The brittle stems are red and the flowers a lovely rose mauve, or sometimes white, and carried in rounded heads in June and July. They persist after their full flowering. A good golden-leaved form, 'Aureum' is particularly attractive in spring when the pretty rounded leaves can light up a pathway edge. As the summer progresses its leaves scorch badly so try to find a place where it is protected from the sun during the middle of the day without being in too much shade. A tiny form, 'Compactum', makes little sweet-smelling plants lovely for window boxes and containers generally, or for a trough of sweet-smelling plants. The leaves are a really bright golden green and the flowers open a little later in the summer than most marjorams.

Small herb garden at Gaulden Manor, Somerset

A less attractive plant, though equally sweet-scented, is the golden-tipped marjoram which has white flowers and golden tips to most of the leaves, but by no means all of them. Regular cutting back will help to ensure new gold-painted leaves.

***Perovskia atriplicifolia*, Russian sage,** is included although it is a semi-shrubby plant, because it adopts the habit of a herbaceous perennial and benefits from being cut right back each spring. Its silver white foliage and downy stems contrast strikingly with the soft blue flowers which are decked along the length of the stem in July and August. Bruise it to release a sage-like aroma. A handsome plant hailing from Afghanistan, it needs a sunny spot in well-drained soil. Cultivars 'Blue Haze' and 'Blue Spire' are both grown, the latter less attractive in foliage form. However, the added bonus of the frosted growth is a lovely attribute and it is very unusual to find a hairy plant that is aromatic.

***Rosmarinus officinalis*, rosemary,** is known for its crisp evergreen spiky foliage that forms a good lax shrub. Gentle powder blue flowers bloom officially in April and May but in many areas and in mild seasons from January onwards. The foliage is strongly aromatic and spicy, oily and distinctive, the shoots always turning upwards at the tips however sprawling the bush becomes. There is a fastigiate form catalogued as 'Miss Jessopp's Upright' with paler flowers. *R. officinalis prostratus* is in fact *R.* × *lavandulaceus*, which spreads mats of foliage speckled with flowers. This one is especially good at the top of a sunny bank or above a small retaining wall where it can be shown off to good effect and at the same time be given the sharp drainage it needs to see it through the winter.

Salvia This is an enormous genus, some of which have aromatic foliage to a greater or less extent and ranging from the blatantly bitter to fragrant. Garden sage, *Salvia officinalis*, is well known, varying in pungency according to the site. Perhaps in the scented garden its purple-leaved and tricolor variegated forms have a place; both are short-lived perennials with a woody base. For conservatory cultivation try *S. grahamii* for its aromatic leaves and scarlet flowers which will continue over a long period during the summer. Propagation of the sages is from seed where it sets, or by cuttings or layering. All of them love a sunny position, and need well drained fertile soil.

***Salvia rutilans* (*S. elegans*)** is a tender perennial of shrubby habit and is known as pineapple sage, because the fragrance of its leaves which surrenders to the merest brush is richly scented of pineapple. It is an interesting plant that will grow out of doors in very mild localities, and then merely for a season, and makes a good house plant or conservatory plant. It does not like full sunshine and needs to be watered regularly. It has beautifully soft-pointed foliage, every shoot suffused with damson purple. Propagate it by division or by summer cuttings.

***Salvia sclarea*, clary or muscatel sage,** is a biennial but is best cultivated as a half hardy annual, starting the seed as early as Christmas. It is finest in its form *turkestanica*. Variously coloured bracts, pink, mauve and blue adorn the flower spike and are particularly enchanting in the evening. A rich oily lemon scent with a distinct tang is perceptible when the plant is growing but as it matures the scent mellows and sweetens, earning it the name of muscatel sage.

Santolina Often overlooked, although popularly cultivated, the santolinas are among the best of sub-shrubs with interesting foliage. They all have bobble or button-like terminal yellow flowers in late summer and love really sun-baked well-drained spots. Use them as a border or driveway edging, or to cover a bank; most respond well to clipping so they can be kept within bounds. Propagation is from summer cuttings.

***Santolina chamaecyparissus*, cotton lavender,** forms a compact woody plant with silver thread-like foliage which is pungently aromatic when rubbed. The smaller form 'Nana' smells pleasanter.

Santolina virens (syn. ***S. rosmarinifolia rosmarinifolia*),** with emerald green foliage, is a short-lived perennial, disappearing quite quickly in many areas, but is successful when cultivated in a container. The heath-like foliage is very slightly oily to touch and therefore adheres to the skin. Its scent is very strongly camphorous and lingering.

Further Reading

The Creative Gardener's Guide to the Scented Garden David Squire (Salamander Books 1986)
The Fragrant Garden Kay N. Sanecki (Batsford 1981)
Herbs and the Fragrant Garden Margaret Brownlow (Darton, Longman & Todd 1957)
Scent in the Garden Frances Perry (Cassell 1992)
Scent in your Garden Stephen Lacey (Frances Lincoln 1991)
Scented Flora of the World Roy Genders (Robert Hale 1977)
The Scented Garden Eleanour Sinclair Rohde (Medici Society 1931)
The Scented Garden Rosemary Verey (Michael Joseph 1981)
Scented Plants Jane Taylor (Ward Lock 1987)

Index

Page numbers in italic type indicate illustrations

Acacia dealbata 26
achillea 9
African marigold 9
alecost 53
Aloysia triphylla 53
Artemisia 54, *55*
auricula 20, 21
Balsamita 53
bee balm 16, *16*
bergenia 9
bergamot 16, *16*
broom 27, 28, *28*
Buddleja 26, 27
butterfly bush 26
Calendula officinalis 9, *10*
Chamaemelum nobile 54, 55
chamomile 54
camphor plant 53
carnation 11, 12
catmint (catnep) *7*, 17, *17*
Cheiranthus cheiri 9, *10*
cherry pie *9*, 14
Chimonanthus praecox 27
Choisya ternata 27
Cistus 53
clary 62
clematis 26
clerodendron 26
Convallaria majalis 11, *13*
costmary 53
cotton lavender 63
cowslip 22
Crambe cordifolia 9
Crocus *42*, 43, 45
cyclamen 43
Cytisus *27*, 28, *28*
dame's violet 14
Daphne 29, *40*
Dianthus 11–14, *13*
Drimys winteri 25
Easter lily 50
elder 25
elephant ears 9
erysimum 9
eucryphia 25
evening primrose 19
fennel 55
Filipendula ulmaria 14
Foeniculum vulgare 55
French marigold 9
Galanthus 43, 45
Galium odoratum 53
Galtonia 43, *44*, 45
garland daphne (flower) 29
Genista fragrans 28
gillyflower 11, 15
Halesia 25
hamamelis 26
hawthorn 25
Helichrysum serotinum 55
heliotrope 26
Heliotropium arborescens (peruvianum) *8*, 14
Hesperis matronalis 14
Himalayan cowslip 22
honeysuckle 30, 31, 32, *32*
hoya 26
Hyacinthus orientalis 46
Hyssopus officinalis 56
Iris 9, 43, 46, 47, *47*
jasmine 26, 29, 30
Jasminum 30, *31*
jonquilla 52
Laburnum 25
lady's maid 54
Lathyrus odoratus 9
Lavandula 56, *56*, 57
lavender *6*, 56, 57
lemon balm 57
lemon verbena 53
Levisticum 53
Ligustrum 25, 30
lilac 38–9
Lilium 48, *48*, 49, *49*, 50, *50*, 51
lily 43, 48–51
lily-of-the-valley 11, *13*
Limnanthes douglasii 15, *15*
Lobularia maritima 9
Lonicera 26, 31, *32*, 33
lovage 53
lupin, tree 9
Lupinus arboreus 9
madonna lily 49
Magnolia 25, *24*
Mahonia x *media* *32*, 33
Malus 25
Mandevilla laxa (suaveolens) 26
marigold 9, *10*
marjoram 60
meadowsweet 14
Melilotus officinalis 15
Melissa officinalis 57, *59*
Mentha 58–9
Mexican orange blossom 27
mimosa 26
mint 58–60
mock orange 33–4
Monarda didyma 16, *16*
Muscari 43, 51
myrtle *5*, 33
Myrrhis 53
Myrtus communis *4*, 33
Narcissus 43, 51, 52
nasturtium 9
Nepeta *7*, 17, *17*
Nicotiana 18, *18*, 19
Oenothera 19, 29
olearia 25
Origanum 60
osmanthus 25
Pancratium 43, 52
peony 9
pennyroyal 59
peppermint 58
Perovskia atriplicifolia 61
Persian lilac (blue Persian jasmine) 38
Philadelphus 34, 38
Phlox paniculata 19, 20
pink 11, 12, *13*, 14, 15
plumbago 26
poached egg plant 15, *15*
polyanthus 20
polygonum 22
Primula 20, 21, *21*, 22
privet 25, 30
queen of the meadows 14
queen's gillyflower 15
rhododendron 25
rock rose 53
Rosa 34–8, *35*, *36*, *39*
rose *6*, 26, 34–8, *35*, *36*, *39*
rosemary 62
Rosmarinus officinalis 62
rue 53
Ruta 53
sage 61–3
Salvia 62
Santolina 63
sarcoccoca 25
snowdrop 43
snowdrop tree 25
sorbus 25
southernwood 54
Spiraea ulmaria 14
stachyurus 26
stephanotis 26
Stewartia serrata 25
stock 15
sweet cicely 53
sweet pea 9
sweet rocket 14
sweet violet 23, *23*
sweet william 11, *13*
Syringa 38–9
Tagetes 9
tobacco plant 18, *18*
trachelospermum 26
Tropaeolum majus 9
Viburnum 41
violet 22, 23
Viola 23, *23*
wallflower 9, *10*, 15
wattle 26
wintersweet 27
Wisteria sinensis 26, *40*, 41
woodruff 53
wormwood 54
yellow melilot 15